BACKYARD SUGARIN' II

BACKYARD SUGARIN' II

By Rink Mann

Photographs by Daniel Wolf

The Countryman Press, Taftsville, Vermont 05073

ISBN 0-914378-20-1

Printed in the United States of America
By The Whitman Press, Lebanon, N.H.

CONTENTS

Cement block, the backbone
of backyard sugarin'

INTRODUCTION

I suppose I got involved in backyard sugarin' the day my determination to make maple syrup ran smack dab into my good wife's determination that the boiling down *not* be done on the kitchen stove. I must say she has a point. You see, the main thing about making maple syrup is you have to boil off about 32 parts of water in the form of steam to end up with one part of maple syrup. That means that if you're boiling down a batch some Saturday afternoon on the kitchen stove and are aiming, say, for 3 quarts of syrup, you're going to put about *24 gallons* of water into the air in the form of steam before the boiling's done. Unless you've got one awful powerful exhaust fan, you end up with water streaming down the walls and enough steam to impair visibility across the room. And, when things finally do clear, you're apt to find the wallpaper lying on the floor. Then too, even if the batch doesn't boil over on you, which it can, the sugar in the spray from all that furious boiling gets all over the stove and is harder than blazes to get off. So, if you want to maintain a measure of domestic tranquility, the best thing is to do your boiling — most of it, anyway — outside, or in a garage or a shed if you've got one handy.

Anyway, the day I lost my kitchen privileges was the day I started figuring out in earnest what I might need to set up a proper evaporator in a little sugar house and get the equipment necessary to do the job right. I was soon up to my eyebrows in catalogues and books on the time-honored equipment and methods used to make maple syrup. This all made good reading, but the smallest evaporator I could find at that time was designed to handle up to 250 buckets, capable of producing about 75 gallons of syrup during the season, and it cost better than $600. When I went to figure out the buckets I'd need to collect enough sap to make it worth while to run the evaporator, plus the holding tanks, instruments and other gear, not to mention building a small sugar house to house everything in, I was looking at an investment well up into four figures. It became clear I'd have to get into the business of *selling* syrup just to make ends meet.

Having other businesses to attend to, I wasn't about to make that kind of commitment to sugarin', but I was just as determined to make my own syrup — say 3-4 gallons a year. I had my own sugar maples, plenty of firewood, an attraction to maple sugar like a bear has to honey, and enough Yankee (or maybe it's Scotch) blood in me to take pride in saving upwards of $12 a gallon in the process.

So, the only solution was to improvise. I scrounged up an old 18″ by 24″ hotel baking pan, built a firebox under it out of cement blocks, with some used stovepipe sticking out the back, and produced a very satisfying batch of golden delicious right out there in the backyard.

That was just the beginning. During the course of the season I ran into, and then started searching out, other backyard operators. We always took time to inspect each other's rigs and to speak kind words about some particularly innovative piece of equipment, be it a rotating bent stovepipe to create a venturi draft effect, or a bathtub holding tank. Naturally we'd steal each other's ideas and make constant modifications in our own rigs during the season, and we'd swap theories on what kinds of maples produced the sweetest sap and what methods should be used to tell when the syrup was ready to be "drawn off."

The real challenge in backyard sugarin' is to find ingenious ways to collect and boil down sap without spending any money, and I must say I found a whole breed of like-minded people. Backyard sugarin' builds interesting friendships, a kind of fraternity, I suppose, born of a mutually parsimonious nature.

I like to think, too, that most backyard sugarers must have a little of the moonshiner's blood in them. And, there *are* a surprising number of similarities between boiling maple sap and distilling out the old mountain dew. In both cases you're separating water from something else. In the case of sugarin' you want what's left in the pan after the boiling, while with moonshining it's what comes off that counts. In both cases, too, you try to set up operations in a nice secluded spot, where you won't get laughed at for your mechanical eccentricities (in the case of sugarin') or arrested (in the case of moonshining).

In an earlier edition of this book, I spent a good deal of time describing, with pictures, some of the more ingenious contraptions being used for boiling down maple sap in the backyard, dealing less thoroughly with other important aspects of backyard sugarin' — like when and how to tap what kinds of maples, and some of the intricacies of boiling down sap without professional equipment. This time around I'm going to try to take the reader through the whole process, step by step, but still emphasizing, each step of the way, how you can get the job done without it costing an arm and a leg.

I am greatly indebted to my wife Louise, who, as I mentioned earlier, was responsible for getting me into backyard sugarin' through her gentle persistence about my not boiling on the kitchen stove, to the several backyard operators who so modestly allowed their sugarin' operations to be photographed, and of course to Dan Wolf, who has captured so well in photographs the essence of the backyard sugarin' industry.

PLANNING AHEAD

With backyard sugarin', one of the things you have to do is plan ahead.

Since the amount and type of equipment you'll need depends a good deal on the amount of syrup you're planning to make, the first thing to decide is just how greedy you are for the golden delicious — that is, how many quarts you want to end up with.

For purposes of demonstration, let's assume you've decided you want 5 gallons, or 20 quarts of syrup by the end of the sugarin' season, and that because you do have other things to tend to during the mud season, which is also sugarin' season, you can only do your boiling on weekends.

With that decision out of the way, we can make the following calculations (with the conviction that we'll get a few arguments about them from other backyarders):

— for each quart of syrup you want you'll need one tap hole in a sugar maple, assuming you collect all the sap and don't spill too much of it on the way to the evaporator. Thus, for 20 quarts, you're going to have to drill 20 holes, set 20 sap spouts and hang 20 buckets. Later on I'm going to argue that a good substitute for a $3 metal sap bucket is a one gallon plastic milk bottle. Therefore, if you accept that idea as a good money-saver, you ought to start saving up at least 20 milk bottles, better 30 or 40 to have ample spares. That requires getting started well in advance of the season.

— the number of maple trees you'll need depends on their size. You're not supposed to tap a tree under 10″ in diameter, but you can put two taps in a tree over 18″ in diameter and three in one that's over, say, 28″. On a real big old maple you can hang even more buckets, but when I see some stately old maple festooned with five or six buckets it makes me think someone's bleeding it to death. So, I try to make do with fewer buckets on more trees. In any case, getting back to that 20 quart plan, you're going to need, as an example, four 3-bucket trees, three 2-bucket trees and a couple of 1-bucket trees, or whatever other combination adds up to 20 taps. The next chapter of this book is going to deal with when, how and where to tap what kind of maples. Suffice it here to say that you should plan ahead, know what maples you'll need to tap and get permission to tap them if they don't belong to you.

— according to the generally accepted rule of thumb a professional wood-fired evaporator will consume a cord of firewood for each 25 gallons of syrup being made. In case you don't know, a cord is a pile of wood measuring 8 ft. long, 4 ft. wide and 4 ft. high. Theoretically, therefore, you'd need 1/5 of a cord to produce your 5 gallons of syrup, but don't you believe it. Since you'll probably boil down on 4-5 separate weekends during the season and will be using a homemade evaporator, which can hardly match the efficiency of the professional rigs, you'll need probably half a cord of good, dry wood to make your 5 gallons of syrup. Now, in boiling sap you want a good, roaring fire, not a slow burner like in your living room fireplace. And, it doesn't make a lot of difference what kind of wood it is just so's it burns well. If you've got to buy your wood, buy hard or soft wood slab (the first slice off the log with the bark still on it), and be sure it's dry, or buy it a year in advance so it will be. Put it near where you plan to set up boiling operations, and put something over it, so your woodpile won't be soaking wet or hardbound with ice and snow when it comes time to use it.

In my own case, being in the real estate business most of the time, I own some wooded lots, and in the first cool days of Autumn I find it quite easy, and enjoyable, to cut down and drag out enough deadwood to more than meet my needs. That's also good woodlot management practice, so that if you don't own your own woods, someone who does would probably welcome your efforts to clean up *his* woods. I've never bought any wood for sugarin'.

Either way, the thing is to figure your firewood needs and get it collected and under cover before the snow flies.

— finally, give some forethought to how you're going to can and store your golden hoard.

For 20 quarts of syrup you'll want to save up enough containers during the year to hold that amount — preferably metal cans, although glass jars with screw-on tops are fine if they don't break when you put the hot syrup in them. My own choice is to save up coffee cans, the ones that come with plastic lids for resealing. The 2 pound size holds a half gallon, so for 5 gallons of syrup you'd need to save up 10 2-lb. coffee cans during the year. Of course you can buy those pretty lithographed cans made specially for syrup, but the half gallon size costs better than 65¢ for the empty can and then only if you buy in lots of 100 from the factory. As I said before, backyard sugarin' is finding ways to make and store syrup without spending any money to speak of.

So, let's sum up the things you ought to be thinking about well in advance if you're aiming to make 5 gallons of syrup.

Save up at least 20 plastic gallon milk bottles or other containers to serve as sap buckets.

Pick out your trees for tapping, and get permission, if necessary, to tap them. You're going to drill 20 holes.

Collect about a half cord of good dry wood, pile it near your planned evaporator site and get it covered over.

Save up 10 2-lb. coffee cans with plastic lids, or something comparable for storing your syrup.

If you figure on making more or less than 5 gallons of syrup, adjust the above calculations accordingly.

There are other preparations that can be made in advance, too, like designing and collecting parts for your homemade evaporator, and perhaps whittling your own sap spouts, but these things can be done over the Winter.

Or, you can do everything at the last minute, if you insist, but I can guarantee you a few frustrations, like having 35 gallons of fresh sap and your wood so wet you can't get the sap boiling, or a nice pot of golden syrup and nothing to store it in. Don't say I didn't warn you.

TREES, TAPS AND SAPS

Figuring out what trees to tap, when to tap them, and just where and how to drill the hole and set the spout are important parts of backyard sugarin'. If you do these things carefully, it will increase the amount and quality of sap you'll get, yield a higher ratio of syrup to sap, which means less boiling, and maybe save you a lot of struggling through deep snow to taps that aren't worth the effort.

Selecting your trees

First off, let's talk trees. Naturally you wouldn't mistake a white birch for a maple, nor would you hang a bucket on a telephone pole (although I've seen that done to get a rise out of the city folks). But, there are four different kinds of maple trees native to the northeastern United States, all of which produce Spring sap flows, can be tapped and will produce maple syrup. These four trees are the Sugar Maple (*acer saccharum*), also called the Hard Maple by furniture makers; the Silver (*or White or Soft*) Maple (*acer saccharinum*); the Red Maple (*acer rubrum*), also known as a Swamp Maple; and the Ash Leafed Maple (*acer negundo*), more popularly known as the Box Elder.

The important fact is that the sap from the Sugar Maple contains about 3% sugar, whereas the saps from the others contains half to two thirds as much. Also, the syrups made from these other saps are darker and less flavorful, so that in total, confining your sugarin' to sugar maples means less boiling and better syrup.

The best way to identify the different maples is by comparing their leaves. This points up the advantage of selecting your trees in the summertime. On page 9 I've included some simple drawings of the leaves of the four trees. The box elder is easily spotted, because each leaf is actually a cluster of 3-5, maybe 7 little leaves instead of a single, multi-lobed leaf. Next easiest to eliminate is the Red Maple. It's leaf has basically three lobes, with maybe another small pair at the bottom, and it is quite serrated (jagged edged). But the most prominent thing about the Red Maple is that it has something red going for it just about all year — reddish outer branches, bright red Fall color and fairly large red buds at the ends of the branches right through the winter.

The more difficult comparison, usually, is between the Silver Maple and the Sugar Maple, because both trees have five-lobed leaves, and they are about the same size. The tipoff is that the Silver Maple's leaves are much more serrated, and there is a definite V sinus between the lobes. The Sugar Maple's leaves have few serrations, and there is a definite U sinus between the lobes, like between your fingers.

If you have to pick your sugar maples after the leaves are down and buried under the snow, there are ways it can be done, but they're too complicated to get bogged down with in this book. Best thing to do is to invite someone over for refreshments who does know. You'll find that sugarin' people are a gregarious bunch, full of free (if sometimes conflicting) advice. If you're still undecided after all that, try to use those big old maples you find rather evenly spaced along the sides of most back roads. It's a sure bet they were planted there generations ago by a farm family that in those days relied on maple sugar for their sweetening needs. Look closely — you may even find the scars of old tapholes long since healed over.

In picking out your Sugar Maples, assuming you've got more than enough to give you the taps you're going to need to meet your production goal, keep these things in mind, too:

— the sap flows best in healthy trees with an abundance of branches. Avoid sparsely limbed trees or ones with a lot of dead branches.

— pick trees that will be the easiest to get to come sugarin' time, when there may be several feet of snow on the ground. That's why the old timers put them along the roads.

— since the sap will flow best on warm, above-freezing days, particularly sunny days, following below-freezing nights, pick trees, if possible, with open southern exposures, (e.g. the north side of east-west roads) so that the warm sun on the trunk of the tree will get things running as soon as possible each morning.

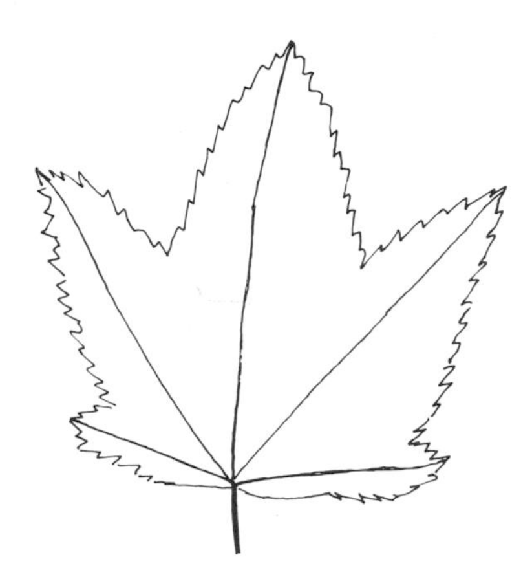

RED MAPLE — 3-5 lobes, well serrated, tapering slightly to ends. Bottom lobes nearly as wide as upper. Definite V sinus. Reddish hues.

ASHED LEAF MAPLE (Box Elder) — 3-7 leaves in clusters. Usually serrated, sometimes not.

SILVER MAPLE — 5 lobes, well serrated, definite V sinus. Silvery undersides. Top lobe tapers somewhat inward at base.

SUGAR MAPLE — 5 lobes, little if any serration, definite U sinus.

When, how and where to tap

Sugarin' time is that time of late Winter, call it early Spring if you wish, when the temperatures drop below freezing at night, but climb well above freezing during the day, pushed up there by a sun which is now climbing higher in the ski, getting up earlier in the morning and lasting longer in the afternoon. The alternate freezing and thawing of the slowly disappearing snow gives it a granular texture, which skiers call corn snow and love to ski on in the warm sun. And, the native folk up country grumble about "mud season," get all worked up over the annual Town Meetin' and start tapping trees for their annual sugarin'. The arguments about just when to tap (combined with sage observations on the weather and past experiences) can be just as heated as the politics. If you tap too early and get an extended cold spell the taps can dry out, and you may have to rebore your holes. Or, if you wait too long, you may miss the first big sap run, which is prized for its high sugar content and fine quality.

I'd say the smart thing to do is to set a spout in one of your trees fairly early, and when you start getting sap flow on a day you figure is fairly typical for that time of year, go ahead and set the rest of the taps.

Here are the things you should pay attention to when you select where to drill the hole, and how to set your spout:

— since the sun's rays speed up the thawing out process each day, you'll get better runs on the southeast to southwest sides of the trunk.

— try to drill directly over a large root or below a large healthy limb since that's where a lot of the sap is headed from and to, and stay away from tap holes of years past.

— if there's a lot of rough bark, it doesn't hurt to knock off an outer piece, so you can get good wood to drill into.

— use a 7/16″ bit to drill the hole. This is the right size to take the standard metal sap spouts you can buy at hardware stores in sugarin' country, or a spout made out of a whittled down 1/2″ dowel with a hole drilled down the center.

— drill your taphole about 1½″ into good wood, which means maybe 2½″ including heavy bark. This isn't a critical measurement. Just be sure you're into good wood, where the sap flows, and have a hole deep enough to accommodate your spout. Drill the hole slightly uphill, so the sap flows outward, and keep turning the drill forward as you pull the bit out, in order to get as many shavings as possible out with the bit. On days the sap is running, you'll see it running out at you almost before you can get the spout in.

— set the spout in the hole and tap it home with a *light* hammer tap (on the front of the metal crest, if you're using a metal spout). The idea here is to set the spout snugly, so it won't fall out of the tree, yet without splitting the maple wood above or below the taphole, where the sap would leak.

— remember that if you're standing on top of a four foot snow bank when making your tap holes at the beginning of the season, those taps may be seven feet off the ground near the end of the season when the snowbank has melted. It's hard to collect sap from buckets 7 feet off the ground.

Buckets and pails

Conventional sugarin' calls for 13 or 16-quart sap buckets made of galvanized steel (in the old days, English tin, and before that wooden buckets). The galvanized buckets come with either flat or bowed covers, to keep rain water and debris out, and they have a hole in the upper edge, where you hang the bucket from a small hook that comes with the metal sap spout. These buckets cost over $3 each (with covers) if you buy them in lots of 100 direct from the manufacturer; more, I suppose, if you have to buy them locally. Since you're going to get enough sap from each bucket to make about one quart of syrup, conventional buckets represent a hefty investment, unless you're ready to commit yourself to their use over and over

again for many years ahead. Most backyarders, therefore, get around this expense by using almost any old container that will hold sap, from peanut butter cans to children's beach pails.

One of the most important developments in the backyard sugarin' industry was the advent of the one gallon plastic milk bottle, known affectionately hereabouts as the Idlenot Dairy Low Fat Sap Bucket (since that happens to be the dairy that supplies our milk). If you're any sort of a milk drinking family,

you can save up more than enough of these gallon milk bottles to provide you with all the sap buckets you'll need. Just rinse them out and hang them out in the garage by a rope threaded through their handles.

The way you hang a low fat sap bucket is to cut a hole with a utility knife about 3/4″ square just below the collar, which is just under the bottle top, and hang the bottle with the sap spout entering this hole and the bottle set down over the back of the metal crest on the spout. Leave the cap on. The bottle will hang there nice as you please, even in a strong March breeze.

Apart from eliminating the high cost of buckets, the Idlenot Dairy Low Fat Sap Bucket has some very real advantages over conventional buckets. For one thing, except for the 3/4″ hole, it is completely enclosed, so you don't get any debris or unwanted predators in the sap. For another thing, it's semi-transparent, so you can see from a distance whether or not it will be worth slogging through the snow to empty it. And, maybe best of all, when the season is over, you don't have to go to all the bother of washing and storing your buckets. You just pitch them into the fire under your last batch of sap. They burn just fine.

The only drawback of the Idlenot Dairy Low Fat Sap Bucket is it's 4-quart size. However, on the kind of days you may have to make more than one collection — when the sun is bright on the snow and warm on your back, and the sap is drip-dripping like crazy — you'll welcome the extra trip back to your sugar maples.

Dealing with frozen sap

Since maple sap flows when it freezes at night and thaws out in the daytime, you can expect to find frozen sap in your buckets if you leave sap in them overnight. Frozen sap presents both problems and opportunities.

The problem comes if you're using milk bottle buckets — the ice won't come out through that small bottle top. The problem is resolved by doing your collecting late in the day. That way there will be very little sap in the bucket to freeze up. If for some reason you forget and find your buckets frozen up solid, have a good extra supply of milk bottle sap buckets, so you can swap empties for the full ones. It'll take the better part of a warm day for the frozen ones to thaw out.

Having extra bottles, incidentally, is a good way to do your collecting. You can carry six empties in and bring out six full ones, empty those in your holding tank and take them back in for six more full ones.

The opportunity in frozen sap is the opportunity to reduce your boiling time, and if you come out some morning and find a good layer of sap ice in your holding tank, you should take advantage of it. The ice actually is very spongy and is made up of frozen crystals of water which have separated out from the sugar, trapping the more sugary sap within the ice, something like a honeycomb. One of the ways the Indians and early settlers used to make syrup was by successive freezing of the sap, each time throwing away the ice, until the remaining liquid was usable as syrup. The problem with this method is that some sugary sap is always thrown out with the ice, so that this method is much less efficient than boiling. In any event, if you do find a couple of inches of ice in your holding tank, prop it up over your holding tank in some way, so that it drains well back into the tank, then pitch it out. You'll have saved yourself some boiling time.

Sap storage

Maple sap, like cider or any fruit juice, can spoil, and care must be taken to keep it as cool as possible and not to store it too long before boiling it down. Sap which has "spoiled" has an unclear, slightly milky look, and if boiled down it will produce a dark syrup. Professional sugarers generally boil down sap within a day or two of its collection, if not the same day. If you're a backyarder, with other things to do during the week and no real desire to sit up all night tending the evaporator, you'll need to figure out not only how much storage capacity you need for the sap collected during the week, but also where to store it with the least chance of spoilage.

Returning to the original example of wanting to end up with 5 gallons of syrup by the end of the season, and assuming now that you will fire up your evaporator for one boildown on each of five successive weekends, the mathematics says you should count on producing one gallon each weekend, for which you will need 33 gallons of sap. Therefore, you will need at least that amount of storage capacity, and it would be wise to have 50 gallons of holding capacity to accommodate heavy sap runs.

A couple of plastic or galvanized ash cans make good holding tanks for an operation of this size, if they can be spared from their normal duties. Naturally they should be cleaned out well, and you can use a plastic liner. Other good holding tank ideas are fifty gallon drums, if you can find a good clean one which hasn't been used for some toxic material, or discarded water pressure tanks, which can be cut open, wire-brushed and painted with a *lead-free* rust proofing paint. I once ran into a discarded bathtub being used as a holding tank. It was strung up, strangely enough, between two trees, like a hammock! Whatever you do use, have a top for it.

Always keep your holding tanks in a location that will be shaded all day long. The idea is to keep the sap as cool as possible, just as though you were dealing with milk. At the beginning of the sugarin' season, when the temperatures are still rather cool day and night, there's little problem, but by late March or April, as the weather warms up, it's something to be mindful of.

Bug filter and holding tank.

HOMEMADE EVAPORATORS

Now that we've disposed of the task of collecting all that slightly foamy, faintly sweet and rather colorless looking sap, the next order of business is to convert it into that gooey, mystically sweet and golden substance called pure maple syrup. The process involves boiling the sap, so that the water in the sap evaporates off in the form of steam, leaving the sugar behind in the boiling pan. Sounds simple, doesn't it, and it really is, although at certain stages of the process, particularly as you're getting your brew close to being syrup, there can be terrifying moments.

Remember, we're talking about starting with, say, 33 gallons of sap and ending up with 1 gallon of syrup. That's about 32 gallons of water to get rid of, which is a lot of steam, which is why wives can get angry at husbands who try boiling sap on the kitchen stove, which is why you're probably better off figuring out a way to do all or most of the boiling outside.

As you know by now, this book deals with how to get the job done without investing a small fortune — in fact, any money to speak of — in evaporating equipment or other fancy apparati. Therefore in this part of the book, we'll examine a selection of homemade backyard evaporators, in the hope that the reader will find, or be inspired to invent, some rig that will satisfy his production requirements and/or meet his aesthetic or mechanical standards.

Engineering principles of backyard evaporators

In boiling down sap, the idea is to get the job done fast by maximizing the amount of steam coming off the surface. You do that in two ways. First, you use an evaporator pan that's relatively shallow (6 to 8 inches) but with a lot of area, so as to have a large boiling surface area relative to the amount of sap in the pan. Second, you design the firebox to get as much flame as possible playing directly on the bottom of the pan. That means the firebox should be relatively shallow, too. And, since the flames tend to be swept backward toward the flue by the draft, rather than upwards against the pan, many backyarders build up the rear bottom of their fireboxes with sand, so that the flames are forced to arch up against the pan, just like with the professional rigs.

The firebox is constructed so that the pan (or pans) sits on the top, supported by the edge of the firebox, but sometimes it's built so the pan can nestle down into the firebox. The reason for nestling is to protect the edges of the pan from the cold breezes, thus hastening the process. Or, if the pan does sit on top, you often find a row of bricks or something, set down along the edge of the pan to serve the same protective purpose.

How big a firebox you'll need depends on how much sap you want to boil down and how much time you can spare to do the job. The more sap you want to boil down in a given period of time, the more surface boiling area you'll need, the bigger the pan (or pans) and the bigger the firebox. (My own new backyard evaporator, which has 864 sq. in. of boiling surface, seems to produce a consistent 1 quart of syrup per hour of boiling time.)

However, there is one thing that should be pointed out. If you plan to boil your sap all the way to syrup on your backyard rig, the pan you end up in should not be so large that your syrup level is too shallow, thus risking scorching your syrup after all that patient work. As an example, our weekender who plans to end up with one gallon should probably not use a pan much over 14″ by 17″, since his syrup will be only one inch deep in a pan of that size (there are 231 cu. in. per gallon).

On the other hand, if he uses an evaporator pan only 14″ by 17″, and assuming for a moment that my own experience is reliable, it would take him close to 14 hours to boil down his 33 gallons of sap.

There are two solutions to this dilemma. Our weekender can use a single large pan to speed up the main boiling and do his final boiling elsewhere (maybe in the kitchen if the wife is gone for the day), or he can use two pans atop his firebox, one, say, 12″ by 16″, which would give him some decent depth for the final boiling, and the other, 16″ by 24″ for boiling fresh sap at the back of the firebox. The combination would give him 576 sq. in. of boiling area, requires a firebox opening of 16″ by 36″, and gives him a 6 hour boil down period, if my own evaporation rates are any indicator.

I put all this down not to get into an argument over whether or not you can make a gallon of syrup from 33 gallons of sap in 6 hours on 576 sq. in. of evaporator pans, but rather to show you the kinds of calculations you have to go through to build a backyard sugarin' rig to meet your needs.

Then too, you may have to approach the problem entirely from the opposite direction and size your rig (and/or how you use it) to a pan or set of pans that are the only ones available to you or already in hand.

Therefore, before we get into specific examples of backyard evaporators, it's appropriate to discuss how you might come by a suitable evaporator pan without having to spend a whole lot of money.

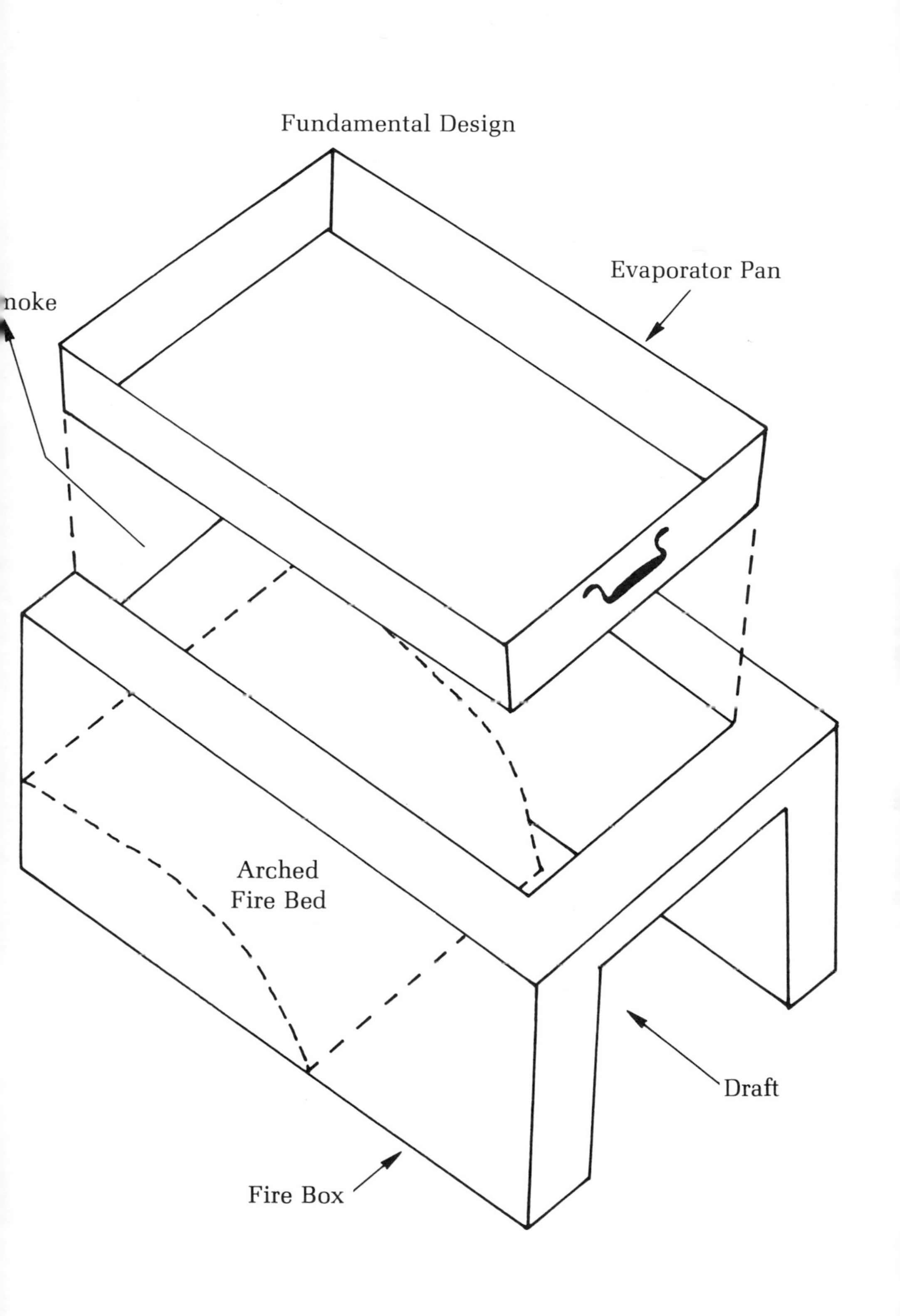
Fundamental Design
Evaporator Pan
Arched
Fire Bed
Draft
Fire Box

Evaporator pans

If all that talk about special size evaporator pans and needing maybe more than one is discouraging you, don't let it. Professionally made pans, with drawing off spigots, fluted bottoms and so forth are frightfully expensive, but you'll find it surprisingly easy to pick up unneeded bake pans at hotels or restaurant auctions that are sized close enough for your needs.

My first evaporator pan was an oversized 18″ by 24″ hotel lasagna pan that nobody wanted. It was covered with "old" lasagna, but after a good cleaning I found I had an excellent steel pan of old-time quality, and it has served me well ever since.

Then too, it's not too hard to make your own pans out of sheet galvanized metal. On the opposite page is a pattern for making a 14″ by 16″ pan out of a 30″ by 32″ sheet of metal. A pan like this, with corners mitered, folded around the ends and locked in place, needs no welding and very little cutting. The bending, particularly working out the corners, takes some doing and a lot of hammering, and while the heavier gauges of metal make better pans, it takes a lot more pounding to get them into shape. The metal for this pan might cost you $5-6.00 in a sheet metal shop.

You shouldn't try to copy the exact design shown here. The best way is to work out your own design with a large piece of brown wrapping paper. You can try folding it different ways until you've got what you want.

Experimenting, improvising, modifying. That's what backyard sugarin' is all about.

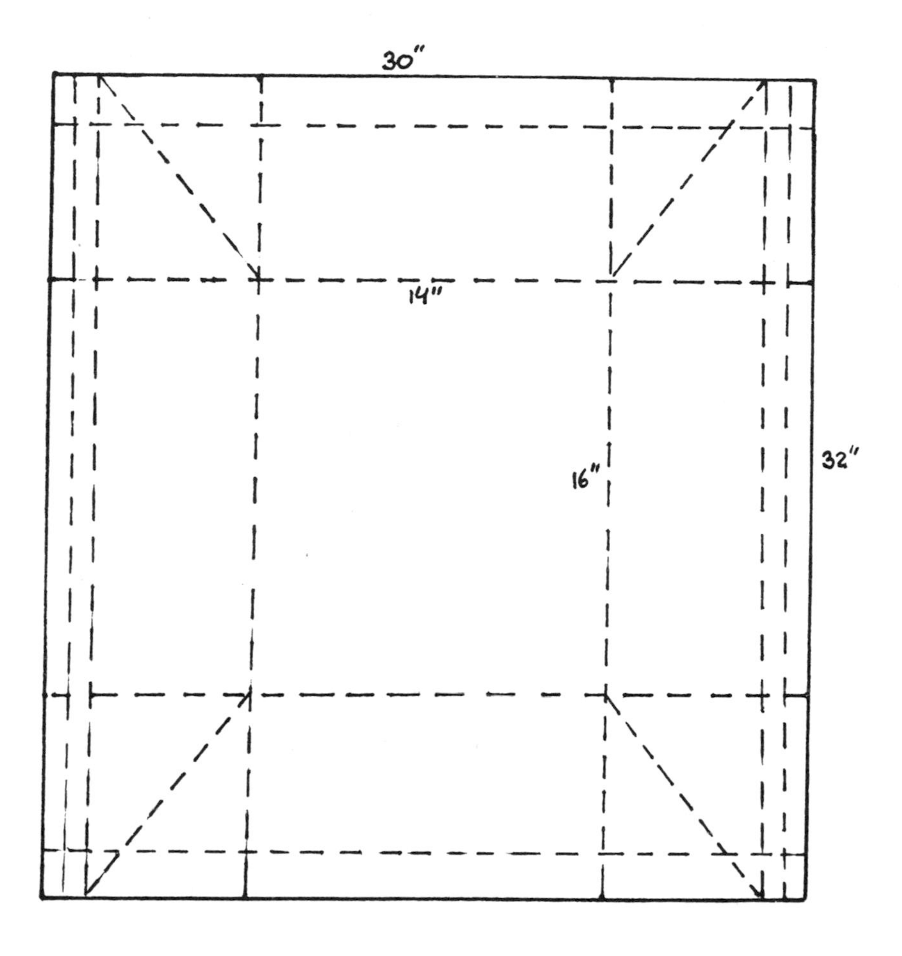
30"
14"
16"
32"

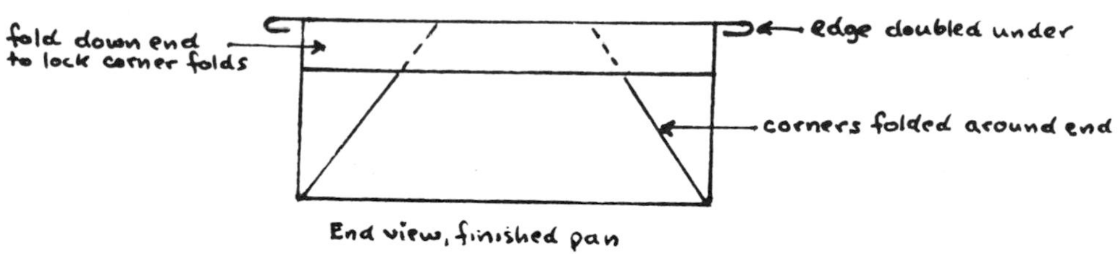
fold down end
to lock corner folds
edge doubled under
corners folded around end
End view, finished pan

Building a homemade backyard evaporator

Making your own backyard evaporator is a process of sizing up what materials you have lying around your place, what you can scrounge up elsewhere, and putting together something that will get a lot of hot flames against the bottom of your evaporator pan or pans with a reasonably efficient use of firewood and with some degree of fire control.

If you are related to a building supply dealer, maybe you can get him to give you some "irregular" cement blocks to build your evaporator out of, or a local plumber might have some discarded water pressure tanks out behind his shop. With an emory blade on your electric saw or a cutting torch, you can make a very professional acting (if not looking) evaporator. Then too, a heating contractor might have an old oil storage tank he'll let you have, which would make up into a nice fire box, and some rusty old stove pipe, which might be just the ticket for your smokestack.

The local dump is well worth canvassing. In my own case I found two very serviceable water pressure tanks for myself, a third for a fellow backyarder, and about 8 ft. of serviceable 8″ stove pipe for my evaporator. Fifty-gallon drums make good evaporator fireboxes, and (if you can find reasonably clean ones) sap storage tanks. I like to think that finding such items in the local dump and putting them back into useful service is a very commendable aspect of backyard sugarin'.

Each backyard sugarer differs not only in his production goals — therefore the size evaporator he will need — but also in the materials available to him and in his skills in making do with what he can get. On the next pages, we'll look at an assortment of backyard sugarin' rigs I've come across, so that the would-be backyarder can pick one that best suits him or, better still, be inspired to invent his own model.

Cement block evaporators

Widely used in the construction of simple, homemade backyard evaporators is the cement block. Construction can range from simple open hearths, laid up without mortar, to more sophisticated set-ups, with various kinds of smokestacks, damper and draft controls.

The cement block is a good building material, because it is symmetrical, not too heavy to work with and permits great latitude in the size and shape of the firebox. This can be important if you have to build the firebox to suit a certain evaporator pan you already have.

Too, the cement block's thickness and interior holes provide good insulating values, holding the heat within the firebox and against the bottom of the pan where it belongs. It is also possible to arrange the blocks in such a way that a chimney of sorts can be made out of cement blocks and be an integral part of the firebox.

The principle disadvantages of cement blocks are that they may have to be purchased (representing an undesirable investment expenditure), and they frequently crack under the extreme heat of the firebox (which may not affect that year's operations but makes it difficult to rebuild the following year).

On the following pages are four examples of backyard sugarin' rigs built with cement blocks. Take your pick — and there are endless other variations.

Basic block sugarin' rig

The rig shown below right is a basic evaporator made of 12 cement blocks set up on a flat, graveled driveway surface, just far enough apart to support the evaporator pan, a reclaimed hotel bake pan. There is no smokestack and no enclosure. A plastic garbage can holding tank is evident, along with a shovel for digging out after snowstorms and a patient friend (background), who adds a protective note to the operation.

Typical of backyard sugarin', fresh sap is continually added to the boiling batch to replace the water boiled off as steam, until the sap is used up. In this instance, rather than pouring cold sap directly into the batch, which might kill the boil, it is poured into a large tin can supported by two rails over the evaporator pan. Here it is partially preheated before entering the pan in a small but constant trickle from a nail hole near the bottom of the can.

Since the fire bed is open at both ends, the flames are carried one way or the other, or both ways, depending on the whims of the wind. Snow must be shoveled all the way around the rig, so that the batch can be tended to regardless of wind direction.

Specifications
Type: Basic Block
Fuel: Misc. scraps of wood and cord wood
No. buckets served: 19
Season production: 22 quarts
Smokestack: none
Enclosure: none
Special feature: automatic sap injection
Owner: Ted Donovan, New London, N.H.

Basic block with stack and damper

Pictured at right is the author's first backyard sugarin' rig, improvised on short notice after loss of kitchen privileges. Essentially a basic block arrangement, a smokestack is added to get better draft and to keep smoke away from the syrup (it can affect taste), and the draft is somewhat controlled by the block and board arrangement at the door to the firebox (foreground) and by the bent license plate in lieu of a stack damper, on top of the stack.

The stack is conventional 6″ stovepipe, set in a discarded 6″ cast iron water main elbow, which happened to be exactly the right size.

It should be noted that in firing up a basic block evaporator, the heat will often thaw out the ground, posing the threat of unstable conditions which in turn might lead to total collapse of the firebox. The picture here shows the author's rig has developed an inward list of about 5° due to this effect, not yet serious enough to imperil the evaporator pan but still a cause for concern.

Specifications
Type: Basic Block with Stack and Damper Controls
Fuel: Owner-cut cord wood
No. buckets served: 14
Season production: 14 quarts
Smokestack: 6″ Stovepipe in Cast Iron Elbow
Enclosure: None
Special features: Arched fire bed and stack controls
Owner: Rink Mann, New London, N.H.

High capacity block evaporator with multiple pans

This attractive backyard sugarin' operation was designed to accommodate a fairly large number of buckets (70) and is an excellent example of self-sufficiency.

The roof, which provides cover against snowfall (and sometimes rainfall), is made of rough poles and plastic and is a veritable school course in truss design. The smokestack is 8″ stovepipe, sized to accommodate the larger fire needed to boil the four pans shown.

The method here used is to add fresh sap to the front pan, where it is brought up to a boil. Then it is ladled progressively to the second, then third and finally the back pan, where the batch is finished.

The multi-pan system has the advantage of adding much boiling surface to the operation and can improve the quality of the syrup by not constantly adding fresh sap to a single pan, a practice that tends to darken the end product.

Specifications
Type: High Capacity Block with Multiple Pans
Fuel: Owner-cut cord wood
No. buckets served: 70
Season production: 40 quarts
Smokestack: 8″ stovepipe with standard elbow
Enclosure: Pole and plastic roof only
Special features: Homemade evaporator pans
Owner: Jonathon Ohler, New London, N.H.

A commendable feature of the Ohler operation is the use of homemade evaporator pans, which would represent a crippling investment if conventional evaporator pans were to be purchased for an operation of this size. Shown below, the pan to the right was cut from galvanized sheeting and soldered by a local artisan. The two pans to the left were folded in a manner similar to that discussed in an earlier section of the book, but with the flaps being held in place with a wooden "handle" at each end, secured with bolts.

Since the Ohler operation serves maples which are further afield, this home made conveyance is used to transport sap from the sugarbush to the evaporator. It's an old pair of skis with a box mounted on top, plus the pushing handle. Here shown with only one collecting tank, the sap sled can accommodate two tanks if necessary. The golden retriever is wondering when someone will figure out how to harness him up to the sled.

Semi-pro cement block rig

If you have the time and are handy with a mason's trowel, you might decide to build a more or less permanent evaporator, like this one, which has an arched fire bed and is made out of cement blocks properly mortared in place. The chimney, too, is of mortared cement block construction. With any mortared structure you must have a solid foundation. Otherwise the frost will heave it up and undo all your fancy mortaring. This one is built on a poured concrete base.

Sitting on top of the rig is a professional evaporator pan, made to order by a leading supplier of professional sugarin' equipment. It's a thing of beauty with it's own spigot so you don't have to ladle or pour the syrup out of the pan (pouring a big pan like this requires steely nerves and a juggler's balance). The only trouble with this pan is that it costs $112.

Note the use of plastic all around the operation to ward off chilling breezes, but the roof here is made of plywood and an old panel door to strengthen it. A good solid roof is a sensible precaution if it's to be a flat one. Heavy wet snows, which are not uncommon during sugarin' time, can cause a plastic roof to sag or collapse.

Specifications

Type: Laid up block arch on concrete base.
Fuel: Owner-cut cord wood
No. buckets served: 106
Season production: 72 quarts
Smokestack: Laid up block integral to arch
Enclosure: Miscellaneous wood and plastic
Special feature: Professional evaporator pan
Owner: Timothy Cook, New London, N.H.

50 gal. and 25 gal. oil drums
make good off-the-ground evaporators

Steel drum and tank evaporators

A growing number of backyarders are building their fire boxes out of discarded steel drums, defunct water pressure tanks and old oil storage tanks, and there's much to be said for this approach. For one thing, a steel drum or tank can be mounted (usually horizontally) up off the ground so that your evaporator pan is at a convenient working level, and you don't have to worry about your fire thawing out the underpinnings. For another, once you've got a firebox that works well, you can store it away each year and not have to rebuild it year after year, as you do with cement block rigs.

50 gal. and 25 gal. oil drums make good fireboxes for single pan evaporators, and the cutting necessary to modify them is not too difficult. However, too much cutting makes them a bit flimsy, and they will rust out sooner than the heavier tanks, especially if you pick one that's already got a few years behind it. Defunct water pressure tanks and old oil storage tanks make excellent fireboxes, particularly for multi-pan operations, and they're rugged, so the cutting comes harder. In either case, tank or drum, the necessary cutting is done with a cutting torch or an emery blade in your circular saw. I've seen oil drums modified by hand with a hacksaw blade, but that's doing it the hard way.

On the next few pages we'll examine five steel drum or tank fireboxes in various styles and sizes. These will no doubt lead the reader to design even more inspirational rigs than the modest efforts shown here.

25 gal. drum evaporator in stone wall

The smallest backyard sugarin' rig I've ever seen was this 25 gal. modified oil drum mounted in a stone wall. The hole for the small evaporator pan, the door and the stovepipe hole were cut by hand with a hacksaw blade.

Mounting the evaporator in a stonewall is a good example of backyarder ingenuity. It gives the evaporator good solid support without having to make special legs, and at the same time it insulates the firebox from excessive heat loss, making for more efficient fuel consumption.

The pan seen to the rear of the evaporator is warming the sap preparatory to adding it to the pan, so as not to kill the boil.

An interesting aspect of this particular operation is that at the end of the season, the smokestack can be lifted for garage storage, and by adding a few rocks to the wall, the firebox is completely hidden from view, neatly concealing it until its unveiling the next season.

Specifications

Type: 25 gal. drum evaporator, wall mounted
Fuel: Sticks and branches
No. buckets served: 9
Season production: 5 quarts
Smokestack: 6″ stovepipe
Enclosure: none
Special feature: Concealable wall mount
Owner: Wilfred S. Davis, New London, N.H.

The mini-drum evaporator shown on the previous page was improvised by Wilfred S. Davis, a retired member of the U.S. Forest Service. Slim's "sugar orchard" is a woodlot next to his house, and his interest in sugarin' is only part of his efforts to bring the woodlot under sound forest management practices.

Sugarin' is a natural outgrowth of good woodlot management, because selective thinning and pruning not only results in better maple sap flows but also provides the fuel for the evaporator — a convenient and efficient way to put the deadwood to productive use.

On the opposite page, Slim shows how he solved his bucket and cover needs. That's an inexpensive plastic paint bucket you can pick up for next to nothing at most hardware stores. The cover has been cut from some leftover asphalt shingles, a good cover idea, because its heavy as well as waterproof and won't blow away.

Plastic paint pail bucket with notched asphalt shingle top.

50 gal. drum evaporator, wood fired

The first time I ran across this rig, I did a double take — figured that stack must have been bent around by a stiff nor'easter. Not so. By putting a bend in the stack and another at the top, and by being able to rotate the stack so that it always leans downwind, this backyarder has created a venturi effect (extra draw) which increases the draft without increasing stack height. Now there's some innovation for you.

This is a neat little rig in other respects, too. A well designed pipe stand holds things steady and well off the ground, for easy wood loading and pan management. And, everything can be taken down easily and stored away until next year.

Specifications

Type: Horizontally mounted 50 gal. drum evaporator
Fuel: Owner-cut cord wood
No. buckets served: 30
Season production: 20 quarts
Smokestack: Bent 6″ stovepipe
Enclosure: none
Special features: Portability and venturi stack
Owner: Tony Dow, New London, N.H.

A closer look at the Dow horizontally mounted 50 gal. drum evaporator shows the combination door and draft regulator — simply the original drum cover hinged at the top. This door can be latched shut, left ajar or propped open in various degrees, to provide a full range of draft control. Note the carrying rails for the evaporator pan. These provide a more level seat for the pan and also add strength to the whole rig. Had these been wider — perhaps simply a portion of the cut out drum folded back — they could also have served to support a row of insulating brick to help protect the pan from the cold air.

Sap injection and preheating are accomplished with a bottom-pierced coffee tin perched on a piece of flat metal at one corner of the evaporator pan. As the can slowly empties, it is refilled with a ladle.

50 gal. drum evaporator, oil fired

Wes Woodward put together this dandy sugarin' rig just outside his garage door back of the house. It's a 50 gal. oil drum set upright, with most (but not all) of the top cut out to accommodate his rectangular pan and a hole in the rear to receive its 6″ stovepipe stack. It's fired, as you can see, with a second hand furnace oil burner. A drum evaporator set up like this would be near ideal if you were using a round washpan evaporator pan.

If you can lay your hands on one, an old oil burner is a convenient way to fire up your evaporator, provides better fire control under the pan, and is probably more economical to operate than a wood fire, if you don't have your own woodlot and have to buy wood. So far as I can tell, too, neither the flavor nor quality of the end product is impaired.

Because of the intensity of the flame from an oil burner and the desirability of keeping the heat concentrated on the pan, this evaporator is lined with fire brick, loosely laid up. Another method for conserving heat is to wrap the firebox in fiberglass building insulation, glass side in.

Specifications
Type: Vertically mounted 50 gal. drum evaporator
Fuel: Home heating oil
No. buckets served: 17
Season production: 28 quarts
Smokestack: 6″ stovepipe
Enclosure: Plywood roof and windshield
Special feature: Lined with fire brick
Owner: Wes Woodward, New London, N.H.

Converted oil storage tank, oil fired

Engineered by the son of a leading heating and plumbing contractor, this evaporator shows the influence of one's vocation on one's avocation. What you're looking at here is half of an oil storage tank, lined with fire brick and topped with an old but authentic professional double-compartmented evaporator pan, complete with a spigot for drawing off the syrup. The rig is fired with a salvaged oil burner, like Wes Woodward's, and the whole operation is enclosed in a board and plastic structure to keep the cold breezes off the pan.

Note the galvanized ash can being used to hold sap. It's worth mentioning that some ash cans are galvanized before fabrication and may leak at the bottom seam. Best check on this before buying one for the purpose. Apparently this one is OK.

Specifications:
Type: Half tank oil fired evaporator
Fuel: Home heating oil
No. buckets served: 30
Season production: 24 quarts
Smokestack: 8″ stovepipe
Enclosure: Board and plastic
Special feature: Finishing rig (see next page)
Owner: Arthur Miller, Elkins, N.H.

CLAYTON A. MILLER, INC.

As boiling sap gets close to the syrup stage, it has a confounding tendency to rise up in the pan, like boiling milk, and boil over unless you can quickly turn down the fire. For this reason, many backyarders prefer to "finish" their syrup in the kitchen or in a separate backyard rig, where they can get better fire control. Here, Arthur Miller is finishing his syrup on his father's plumber's stove (used for melting lead). It works fine, thank you.

Double pan, converted water pressure tank evaporator

This backyard sugarin' rig, recently assembled by the author and fired with wood, utilizes two defunct water pressure tanks, one as a sap storage tank (the one suspended between the trees), the other as a firebox for two 18″ by 24″ evaporator pans. The firebox is supported by two pieces of pipe suspended between two sets of cement blocks, and the whole operation is under a sloping roof made of two overlapping sections of corrugated steel roofing found, along with the pressure tanks, at the local dump. The only out-of-pocket costs involved were the purchase of a sheet of galvanized metal to fabricate the rear pan, a small payment to the mechanic who did the necessary cutting of the tanks with his torch (I could have done it better and cheaper by buying an emery blade for my circular saw), and purchase of the valve under the storage tank and the two hinges for the front door. Even the 8″ stovepipe stack materialized out of the dump, complete with a very important looking but unnecessary backdraft preventer.

This backyard rig, capable of producing about one quart of syrup per hour of boiling, will be used in the following section of this book to take the prospective backyarder, step by step, through the boiling down process. Therefore we will hold back on some of the more interesting details of this operation until that time, showing it here only to properly catalogue it among the others and to touch on a few construction details.

Specifications

Type: Double pan, converted water pressure tank evaporator
Fuel: Owner-cut cord wood
No. buckets served: 24
Season production: 20 quarts
Smokestack: 8″ stovepipe with stack damper
Enclosure: Sloping corrugated roof only
Special feature: Gravity sap feed from storage tank
Owner: Rink Mann, New London, N.H.

This view of the converted water pressure tank evaporator shows more clearly the various cuts that were necessary to provide a front door, an 8″ hole in the rear for the stack and the two large top openings sized to accommodate the two pans. Note also how sand has been used at the rear of the firebox, to force the flames to arch up and against the bottom of the rear pan.

A strip of ordinary fiberglass building insulation has been placed along each side of the firebox, to prevent heat loss, being secured there by three strands of wire running around the firebox.

The two green hardwood poles supporting the corrugated roof sections were slung between four trees with two pairs of worn out tire chains. The chains are tightened around the tree with their regular draw-fasteners, so they will not slide down the tree, and the poles are inserted in one or two of the loops of the chains, as shown here. It proved entirely unnecessary to use any nails to support the roof.

Backyard barbecue conversion

Improvisation, again, is the key to successful backyard sugarin'. For instance, if you already have some sort of outdoor cooking facility, don't overlook the possibility of putting it to work for you, making maple syrup.

Here, backyard sugarer H. Sumner Stanley is beating the high cost of living by happily, and cleverly, making his own syrup on the family barbecue, excavated from a snowbank for the occasion. During a typical season Mr. Stanley produces about 6 quarts of syrup from maybe 7 taps at virtually no out-of-pocket cost and with a lot of fun and satisfaction in the bargain.

Note the usual trappings of the backyarder — sap warming in a white pail forward of the evaporator pan and a bottom-pierced sap injection can on the upper level.

Specifications
Type: Backyard barbecue conversion
Fuel: Owner-cut cord wood
No. buckets served: About 7
Season production: 6 quarts
Smokestack: Integral to structure
Enclosure: none
Special feature: Old wood stove built into stone barbecue
Owner: H. Sumner Stanley, New London, N.H.

Congressman's box stove conversion

Here's another good example of how to make do, come sugarin' time, with an already existing fire box. In this case it's an old wood burning box stove, set up more or less permanently in one corner of the garage, and that's our Congressman, Jim Cleveland, and his daughter, Susan, drawing off a batch of syrup. Jim likes to tend to his syrup-making personally whenever he's not too busy tending to the problems of his New Hampshire constituents.

Jim has had an evaporator pan made that fits just right over the stove opening after the stove lids and other top parts are removed, and the pan has a handy drawing-off spigot.

The 6″ stovepipe stack finds its way into a nearby chimney, which draws a good draft for the stove.

Specifications

Type: Converted box stove evaporator
Fuel: Owner-cut cord wood
No. buckets served: 30
Season production: 32 quarts
Smokestack: 6″ stovepipe into a chimney thimble
Enclosure: Garage
Special feature: Custom evaporator pan
Owner: James C. Cleveland, New London, N.H.

THE BOILING DOWN PROCESS

Having discussed the selection and tapping of trees, the collecting and storage of sap and the building of a homemade evaporator — all of these without having to spend a small fortune on equipment — we now come to the important process of converting all that watery, rather tasteless sap into golden, delectable maple syrup.

I think it might be well, right off, to confess that given the same stand of sugar maples it is unlikely that the backyard sugarer is going to end up with as high a "grade" of syrup as the commercial operator. The main reason is that a better grade of syrup can be produced by boiling the sap when it is as fresh as possible and boiling it down as fast as possible. Unlike the backyarder, who may have other things to do, the commercial operator by and large collects his sap daily or oftener, gets it into the evaporator just as soon as he's got enough sap to start a batch and boils it down fast by using professionally designed fire boxes under fluted, compartmented evaporator pans. Then too, with compartmented pans, the boiling sap moves automatically from one compartment to the next as it thickens and is not being repeatedly diluted and reboiled by the addition of fresh sap, as with a single pan backyard rig. Such practice, I am told, tends to darken the syrup.

At the same time it should be remembered that maple syrup "grades" are established by state laws only to standardize various grades of syrup to be sold to the public, the commendable objective being to see to it that when John Q. Public buys a can of "Grade A" syrup (which he cannot see through the tin) he will be reasonably sure that he's getting what he thinks he is getting. Furthermore, it is not necessarily true that a "Fancy" grade of syrup will be preferred by all, or even a majority of your average maple syrup addicts over, say, the next lower grade, which is Grade A, or even Grade B. Truth of the matter is I've run into quite a few of the professionals who prefer the lower grades, with a darker amber color and a heftier taste.

Anyway, the point is that with backyard sugarin' if you don't plan to sell any of your syrup, you don't have to figure out its grade and label it accordingly, and the only judge of its perfection is the backyarder himself and his family, who are the ones that are going to consume it. I have yet to run into any backyarder whose family had rated his production run of syrup at less than "super!," "yummy" or "hmmmm, good," which are all top grades in backyard sugarin'.

Next thing I'd like to say is that you don't need a hydrometer, or any special instruments in fact, to make excellent maple syrup, unless, again, you plan to sell it. An ordinary candy thermometer is useful, but you can even get by without one of those if needs be.

When sap starts to boil, it looks very much like water boiling. Then, as the water is evaporated off, the color gradually darkens, and as the batch approaches the syrup point the bubbles get smaller and smaller (like milk boiling). Then suddenly, if you don't keep a sharp lookout, the batch can rise up in the pan (again, like milk) and boil clean over. At this near-syrup stage, too, if you dip a spatula into the batch and take it out and the residue tends to come off in small gooey sheets rather than in drops, you've got syrup. Now, all of these characteristics I've just described are things that can be observed without any instruments at all, and experienced backyarders can make fine syrup by visual observation alone.

But, as I said, a good candy thermometer can be useful, because as sap boils and gradually thickens, its temperature goes up due to the increased sugar content. Thus, its temperature is a useful measurement of its progress toward, and arrival at the syrup stage.

With the thermometer what you wait for is that point when the temperature of your batch has risen 7° above the temperature of boiling water on that same day at that same place. I say same day same place, because boiling temperatures vary with changes in barometric pressure, which in turn vary with changes in the weather and altitude. For instance, the temperature of boiling water is 1° less for each 550 ft. of altitude, so if you're 1100 ft. above sea level water boils not at 212° but at 210°. The thing to do is put your thermometer in some boiling water and see what the reading is, then add 7° to that to get the reading you're aiming for.

In practice, the temperature of boiling sap rises very slowly at first and very quickly toward the end of the boil-down. The reason for this is that the concentration of syrup (percentage syrup) accelerates toward the end as there is less and less water to boil off. If you're dealing with an 8 hour boil-down, for instance, it may take 3-4 hours for the temperature to work up 1°, particularly if you're continually adding new sap, but when you're near syrup stage the temperature can rise a degree in a very few minutes. This acceleration and more rapid temperature rise near the end is what makes the thermometer a useful thing to have around.

But, enough of that. A good way to cover the boiling down process might be to take you right through it, step by step, using my own two-pan, converted water pressure tank evaporator as an example. Naturally, you'll run into your own problems and crises with a rig of your own design, but I hope that by looking over my shoulder you'll avoid some of the more obvious pitfalls and perhaps see an improvement you might care to pass along to me.

Step by step on the author's rig

Having collected enough sap in the storage tank (about 50 gals.) to insure ending up with no less than 6 quarts of syrup in the front pan (to avoid scorching in my 18″ by 24″ pan), I turn on the holding tank faucet full blast and get a good two inches of sap in both pans *before* starting the fire. The holding tank is mounted higher than the evaporator pans so the sap can flow by gravity through the short garden hose, which can be suspended over either pan by a string from the roof. Once the initial quantity of sap has been placed in both pans, the sap flow from the tank is valved down to a trickle, and the hose is left over the rear pan. The rate of the trickle should be set to maintain an approximate 2″ level in the back pan, as water is boiled off and to replace boiling sap I ladle up to the front pan to maintain *its* level.

Here, the sap has just started to boil.

Here's a closeup of the hose delivering fresh sap to the back pan. The cloth wired over the end catches any debris that comes along. There shouldn't be much to catch if you're using Idlenot Dairy Low Fat Sap Buckets and keeping your holding tank covered. Too, most backyarders pour their gathered sap into the holding tank through a cheesecloth filter, which is left over the tank to keep predators (bugs) out.

Since the two pans are not interconnected, like progressive evaporator compartments in professional equipment, I tried to figure out a way to siphon sap from the back to front pan automatically. It didn't work. The steam kept getting up into the siphon, breaking up the siphon effect. Here I am, ladling from back to front, as needed, to keep a good level of boiling sap in the front pan.

Since the coffee can I used to ladle boiling sap got a bit too hot to handle barehanded, I made up a little handle for it, which may be noteworthy. It's a piece of green birch, being a small branch coming out from a main stem. If you split the branch in toward the stem, you can slide the can into the split, as shown here, and it will stay there without fastening. At least it did for me.

During the boiling you'll notice some scummy looking foam building up around the edges of your pans. This is a natural byproduct of the boiling and should not cause any alarm. Since it tends to reduce the boiling surface area, though, it should be skimmed off from time to time. I use a kitchen strainer and simply pitch the foam off into the snow. Any residue foam in the batch will be removed when you filter your syrup.

As the boiling proceeds, and I am continuing to ladle boiling sap from the rear to the front pan, I finally run out of sap, first in the tank and then in the rear pan. Since I cannot remove the rear pan (doing so would leave a big hole in the fire-box, ruining the draft), I simply shovel snow into the back pan when the sap level gets too low for comfort. The snow melts down and gives me a handy source of warm (then hot) water for rinsing my filter felts and other pieces of equipment. If there's no snow around, you should have some plain water handy in a bucket for these purposes.

Keeping a weather eye on the lowering depth of syrup in the front pan, the character of the bubbles (they're beginning to get smaller, like boiling milk) and the thermometer (it's now close to 7° over boiling water temperature), I begin to let the fire die — maybe push some of the fire toward the back, open the door to let cool air into the firebox or even remove an excess log or two. My plan is to get as close to finished syrup as possible without the batch boiling over, then ladle and pour it into waiting half gallon (2 lb.) coffee tins.

All of a sudden (and I mean sudden), the foaming bubbles in the syrup come together, and the whole batch starts rising up in the pan, and it appears that it will boil over, a major crisis in backyard sugarin'. Acting clear-headedly and swiftly, my solution here was to shovel snow directly into the firebox. It works fine, particularly if you're finished boiling anyway. Another solution — the time-honored one — is to touch the surface of the boiling with a bit of butter on a stick, or a piece of bacon. Like oil on troubled waters, the boiling magically subsides. It works.

Here I am, ladling the day's production run into coffee tins, filtering it as I do so. The syrup has something in it called niter, which would settle out at the bottom of the can, make your syrup cloudy and maybe give it a bitter taste if not filtered out. In my early sugarin' days I lined a kitchen strainer with water resistant paper towels. They worked fine, but the filtering took forever. So, I acquired a piece of regular filtering felt (a commercial operator friend of mine gave it to me), and I must say it's far superior to paper toweling, and of course it can be rinsed out (don't launder it with detergents) and be used over again many times. Before filtering hot syrup through a felt filter you have to moisten it first with hot water, and you can rinse it out between pourings to get rid of accumulated niter.

When the syrup level in the front pan gets too low to ladle conveniently, I simply pick up the whole pan and pour directly from it. Don't try that too soon. Pouring an inch of hot syrup from an 18″ by 24″ pan takes a real juggler.

Well, that wasn't too harrowing, was it? So now I have six two-pound coffee cans full of red hot syrup, and I'm not exactly sure whether its underboiled, just right, or overboiled. What I do is put the plastic lids on the syrup and head for the kitchen. There I bring the syrup back up to a boil in a large pot and check it again carefully with my thermometer and with a spatula for the aproning effect.

To suit my own taste for a slightly thicker syrup I make sure the temperature is a strong 7° to 7½° over boiling water temperature (boil some water on the next burner to check your thermometer's present reading for boiling water). The aproning effect gives you a feeling for the syrup's consistency, and traditionally, syrup is not syrup until it aprons. Dip the spatula in the boiling syrup and lift it straight up. If the syrup on the spatula drips off in regular drops, keep boiling. When the drops begin to hold together and the syrup tends to slide off in little sheets (or aprons), the batch is ready for canning.

Incidentally, if you have overboiled put some boiling water in the batch, until it is back down where you want it. Otherwise sugar crystals may form in the syrup cans. If you underboil the syrup will not only feel watery but will be more susceptible to spoilage.

This is what syrup looks like when it's about to boil over. Quick! Someone lift the pot off the burner (or touch the syrup with a bit of butter).

Canning and storage

Canning maple syrup presents the same problems as the hot canning of cooked vegetables, with at least one important (and happy) difference. If a jar of canned tomatoes goes bad, you've had it. With syrup, if it gets moldy, you can scoop off the mold, bring the syrup back up to a boil, and you're back in business.

Commercial operators filter their syrup through felt filter cones into filter tanks and then valve the syrup directly into clean, new syrup cans while the temperature is still not less than 180°. The cans are filled right to the top, so there'll be a minimum of possibly contaminated air, the caps are put on, and the can is sloshed or set on its side, so that the hot syrup sterilizes what air there is. The syrup, of course, after all that boiling, is about as sterile as it can be. Syrup packed this way should keep very well on the pantry shelf, but like with canned vegetables a cool dry place is better.

For home packing of syrup I prefer 1 lb. and 2 lb. coffee cans with plastic lids (they come on most coffee cans these days for resealing). You can save up what you need during the year, they're very clean (because they've contained a dry product) and you can get a good seal (by using freezer tape around the plastic top after it is put on). But you can use anything for storing your syrup, provided it is clean (really clean), won't break when you pour hot syrup into it and can be sealed. Glass jars with screw-on tops are fine if the seal in the top is still intact.

In my case, while the syrup is being reboiled and checked on the stove I thoroughly rinse out the cans. Then, when the syrup is ready, I pour it immediately into the cans, again through my felt filter, put on the plastic lids, seal with freezer tape, slosh a bit to sterilize the lids and set aside to cool. We know the seal is good when we see the plastic lid drawn down tight in a concave shape as the small amount of air in the can cools and contracts.

I'm sure that syrup packed this way would not spoil, and we have never had any spoilage, although I must confess that we store our syrup in the bottom of our upright freezer.

Making maple sugar (for those who dare)

To tell you the truth, I've never made any maple sugar or maple cream, unless you count the time some years ago, when my attention got diverted during a kitchen boil-down, and my syrup was reduced to something resembling road tar. I managed to get it out of the evaporator pan and into another pan before it hardened up on me, which it did while I was looking the other way. The spoon I had been using became permanently lodged in the pan, as well as the sugar. Having a prankish streak in me, I wrapped up the whole embarrassing mess in some brown paper, spoon, pan and all, and mailed it to a friend of mine with the announcement that it was a first production run of *crême brulé*. I haven't heard from him since.

The moral of the story, however, is that when you start dealing with syrup boiling at high temperatures, things can happen fast, and you have to make quick, sure moves at the right time.

Making maple sugars of different types is simply a matter of boiling your syrup beyond the syrup stage, to various temperatures, then doing certain things quickly to achieve the desired end product. For instance, remembering that syrup is syrup at about 7° above boiling water temperature:

— At 25-27° above boiling water temperature, you can pour your batch out on a cooling slab, to get it down to room temperature as quickly as possible without any agitation, then whip it vigorously until you end up with that creamy, delectable confection known as maple cream.

— At 30-33° above boiling water temperature you can stop boiling, let the batch cool naturally and undisturbed to just below boiling water temperature (careful not to stir it with the thermometer), at which point stir vigorously. All of a sudden it will start to crystallize (says my source), so you have to move fast and get it in whatever molds you use before it hardens up in the pan. This method gives you standard maple sugar.

If you go much higher than 35° above boiling water temperature, you're on your own. You're apt to end up with sugar you can only break up with a hammer, or with another batch of *crême brulé.*

For more detail on how to make hard maple products, I'd recommend that you check your library for books on the subject. There are some good ones, written by people who've had a lot more experience than I have in this intricate art.

MUSINGS OF A BACKYARD SUGARER

I don't want to get nostalgic or overly philosophical about sugarin', because it really is a fairly simple, down to earth and practical procedure which yields a very useful product which almost everybody enjoys eating. But I've got to say that there is something magical about sugarin', and if you talk with people who make maple syrup, either in a big commercial evaporator or out in the backyard, you'll find out there's a lot of agreement on that fact.

Maybe it's the time of year — the warm sun climbing higher into the sky, warming the back after a long winter, turning the snow to piles of white corn, turning the brooks from trickles to torrents, starting the maple sap flowing — a sort of hint of the Spring and Summer lying ahead. Maybe it's the drip drip of sap falling into the buckets, the telltale aroma of boiling sap or the hissing sound of sap in a rolling boil. Maybe it's simply the magic of converting sweet water, as the Indians used to call it, to delicious golden syrup. But whatever it is, it's there.

Mix the magic with a liberal dose of ingenuity, mechanical innovation and the determination to make do with materials at hand, and you come up with what this book is all about, backyard sugarin'. In one fell swoop you can satisfy your creative instincts, indulge yourself in a mystical experience and fill the pantry shelves with a product that the whole family can enjoy at a fraction of its usual cost.

Well, as I write these words, another season has just ended, my sap spouts are stored away (that's the only equipment I need clean and store from year to year), and my head is once again full of ideas about building another homemade evaporator for next year. I want it to be a bit smaller, as I don't need as much syrup as my water pressure tank evaporator is capable of producing, and the front pan will be somewhat smaller and maybe deeper, so it will be easier to handle. Maybe I'll figure out a way to interconnect the two pans. Then too, I mustn't forget to mark those three big maples that produced such sweet sap (and so reliably!).

But there I go again, never satisfied to leave well enough alone. Maybe I should just be content with the memory of that warm, sunny afternoon, not so long ago, when a bottle of ale lay in the corn snow, awaiting my indulgence, and the boiling sap hissed quietly in the background. And just think of it, there wasn't a customer in sight.